Mountain Night
Mountain Day

by Anthony D. Fredericks

illustrated by Kenneth J. Spengler

Rio Chico

Books for Children

Can you find who is hiding in the shadows?

In every illustration, look for a glimpse of the animal from the previous page and a hint of the animal from the next page. Don't forget to spot the mouse, too!

Mountain night,
Dancing light,
Critters waken,
Full moon bright.

Scatter, scurry,
    Black, white, furry.
Watch out! Skunk—
    In a hurry.

Above the ground,
Not a sound.
Great Horned Owl
Looks around.

Up in a tree,
Silently,
A small gray fox.
Can you see?

Seldom sleeping,
Slowly creeping,
Mountain lion.
Look! She's leaping.

Blackened eyes,
Very wise.
Sly raccoon
Pulls and pries.

Over there,
Big black bear,
Nursing babies
In her lair.

Now night is through.
Morning dew,
Sparkling meadows,
Skies of blue.

Mighty, strong,
Antlers long.
Majestic elk
Sings his song.

Hill and dale,
Forest trail.
Baby deer,
Wagging tail.

Chatter, squeak,
Dart and peek.
Active squirrels
Along the creek.

With long, broad wings
By lakes and springs,
Majestic eagles
Fly like kings.

Darting, hiding,
Rapid-riding.
Rainbow trout,
Swimming, gliding.

A fallen tree,
Lakeside, see!
Baby turtles,
One, two, three.

Sun, moon high, a mountain sky,

Busy creatures crawl, run, fly.

Field Notes

The western United States is one of the most mountainous regions in the world. Several mountain ranges, including the Rocky Mountains, the Sierra Nevada Mountains, and the Cascade Mountains are located throughout the western states. These mountains are known for their snow-capped peaks, rushing rivers, crystal-blue lakes, deep forests, and sparkling waterfalls.

Many different animals make this wild and wonderful land their home. Some are nocturnal (nock-tur-nal). They sleep during the day and are active at night. Other mountain critters are diurnal (die-ur-nal). Like you, they are awake during the day and sleep at night.

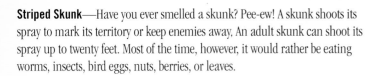

Nocturnal Animals
[NIGHT TIME]

Striped Skunk—Have you ever smelled a skunk? Pee-ew! A skunk shoots its spray to mark its territory or keep enemies away. An adult skunk can shoot its spray up to twenty feet. Most of the time, however, it would rather be eating worms, insects, bird eggs, nuts, berries, or leaves.

Great Horned Owl—You may have heard this owl when camping or traveling through the mountains: *hoo hoo hoooo.* It is a powerful hunter and will attack many small animals such as rabbits, mice, snakes, and sometimes skunks. It has very sharp talons and a sharp beak to tear apart its food.

Gray Fox—Believe it or not, the gray fox likes to spend a lot of time in trees, where it hides from its enemies, chases birds, or steals eggs from nests. Fox parents will sometimes build their dens high in trees to keep their young ones safe from danger.

Mountain Lion—A mountain lion is a large unspotted cat. It is a very good climber and jumper, and can leap more than twenty feet in a single bound. It is often the top predator, or hunter, in the mountains.

Raccoon—A raccoon is one of the smartest creatures around. It is known for climbing over fences and breaking into garbage cans. Its black eyes make it look like a bandit scurrying through the night.

Black Bear—A black bear can range in color from cinnamon to black, and it often has a white blaze on its chest. A fully grown black bear can get quite large, frequently reaching six feet in length and 585 pounds in weight. Adult females give birth to a litter of one to five cubs (typically two) in January or February.

Diurnal Animals
[DAY TIME]

Elk—Herds of elk can be found in mountain pastures or wooded areas throughout the western states. An elk can grow to five feet in height and may weigh up to 1,000 pounds. Males often have large branching antlers; females do not have antlers. Adult males make a delightful bugle-like call that travels long distances through the mountains.

Whitetail Deer—The first thing you notice about this animal is its white tail, which can be seen as the deer disappears into a nearby forest. When a mother deer is feeding, it will often leave its baby hidden in tall grass to hide it from potential enemies. Whitetail deer live primarily in deep forests and may form herds of up to 150 members.

Red Squirrel—This mountain critter makes a lot of noise. It chatters, squeaks, and makes other loud sounds when an intruder enters its territory. It loves to eat the seeds from pinecones in much the same way you eat corn on the cob. Its fur is rusty-brown and its tail is white underneath.

Golden Eagle—This majestic and graceful bird lives throughout the western states. Its acrobatic displays are truly a sight to see. It nests on steep rocky cliffs and hunts its prey in deep canyons and along soaring mountain peaks. It is about three feet long, but its wingspan can be more than six feet!

Rainbow Trout—Found throughout mountain lakes, rivers, and streams, this beautiful fish gets its name from the reds, pinks, browns, and greens that cover its body. Its upper body has a broad patch of speckles (like freckles). It's an important food for other alpine creatures such as otters and minks.

Western Painted Turtle—You can find this turtle basking on logs or rocks near ponds or lakes. If you approach, it will slip into the water and disappear. Its shell is commonly olive green, black, or brown. It gets its name from the bands of red, yellow, and green on its underside.

To Katie Beauchat
for her constant and enthusiastic celebration of "read-alouds."
—A. D. F.

To my nephews and nieces:
Ryan, David, Melanie, Nicole, Mikey, and Andrew!
—K. J. S.

Rio Chico, an imprint of Rio Nuevo Publishers®
P. O. Box 5250, Tucson, AZ 85703-0250
(520) 623-9558, www.rionuevo.com

Editorial: Theresa Howell
Book design: David Jenney

Printed in China.

10 9 8 7 6 5   18 19 20 21 22 23

Library of Congress Cataloging-in-Publication Data

Fredericks, Anthony D.
  Mountain night, mountain day / by Anthony D. Fredericks ;
illustrated by Kenneth J. Spengler.
     pages cm
  ISBN 978-1-933855-98-1
 1. Mountain ecology—Juvenile literature. 2. Mountain animals—Juvenile literature.
I. Spengler, Kenneth, illustrator II. Title.
  QH541.5.M65F74 2014
  577.5'3--dc23
                    2013025358